I0079622

Pigsticking in Morocco and India

By

Mrs Mansel-Pleydell
&
R D Rudolf

MRS. MANSEL-PLEYDELL ON NIZAM

PIGSTICKING IN MOROCCO

BY MRS. MANSEL-PLEYDELL

I HOPE I am not unduly flattering myself in believing it to be possible that readers of the *Badminton* may be interested in my experiences of pigsticking near Tangier, in Morocco, as I was fortunate enough to kill, single-handed, the finest boar (with one exception) that has ever been speared in the annals of the Tangier Tent Club; and, indeed, according to Colonel Baden-Powell's well-known book on " Pigsticking," a boar so large that its equal in size is rare even in India. The pigsticking takes place in a tract of country of about five miles radius, the sporting rights of which were given over to the club by the present Sultan's father. Sir John Drummond Hay, then British Minister in Tangier, having obtained this concession from the Sultan, inaugurated pigsticking for the benefit of his countrymen with sporting tastes who might find themselves resident in Morocco. The camp is usually pitched close to a Moorish village called Hawara, about fourteen miles from Tangier, and fairly central for all the meets. The country we hunt over is for the most part extremely thick, as it is extensively covered

1

with dense bushes and scrub, which in some places have grown so high that it is exceedingly difficult to keep a pig in view; nor can it be described as anything like good going, those who are unaccustomed to it finding it very hard to keep in their saddles when riding over it at a gallop, as the horses dodge cunningly among the bushes and constantly jump them when least expected. On the other side of our camp is a large mud flat which must in recent years have been reclaimed from the sea. This mud lake running up between two thick coverts makes a splendid place to drive the pig across ; many are the good runs that have been enjoyed there, and many fine boars have fought for their lives and died there, bravely, in the open. However, at this meeting not one broke across; no doubt they considered discretion the better part of valour, as the mud was so dry that a pig's chance of escape from a fast horse would have been small. There is another open piece of ground bordering the cork woods which lies between Hawara and Tangier. This is a large marsh, which always contains more or less water, according to the seasons, and is in some parts very heavy going. It was there that I fell in with my "record boar."

As a rule we have four meetings each season, but last year had only been able to have one, in November, when owing to bad luck and a lot of rain no more than two boars were killed. A second meeting was arranged for the first week in January, but had to be twice postponed on account of the rebellion among the local tribes ; for most of our beaters, some fifty Moors all told, were involved, This made it impossible for us to hunt, more especially as the head of the beaters was one of the leading rebels. But rebels and cutthroats though undoubtedly many of them are, they are the keenest of sportsmen, and when a "haloof" is found their excitement is intense. They shout, fire their long guns as a signal that the game is afoot, and do not hesitate frankly and freely to express their disgust if, after the boar breaks, he escapes untouched. They are very superstitious also, and invariably find a reason when we have bad luck, generally pitching upon one of the spears as being the "Jonah, ' and are quite convinced amongst themselves that we shall have no sport as long as we harbour the wholly unsuspecting individual whom they are pleased to regard as the source of misfortune. The "boar hounds" are a very queer-looking lot. They are just like the pariahs which haunt an Indian village ; but, unpromising as may be the appearance of the pack, their looks in most cases belie them, as there are usually two or three on the track of any pig afoot. We generally lose several hounds each meeting from being ripped, so they might well be worse than they are, and want of pluck is at any rate not their failing.

PIGSTICKING IN MOROCCO

After a period of exciting warfare about which there was a great deal of noise and very little result, things quieted down; we were able to fix our second meeting for February 2, and we went out into camp, a party of five men (including my husband, who is manager) and myself representing "the spears," with three others who came out as onlookers. The "spears" were augmented each morning by two, three, or four extra men who rode out for the day's sport, returning to Tangier when it was over; so we were, as a rule, nine or ten spears.

We had very good sport each day. On the Monday two pigs were killed. One, a fine boar, gave us no little excitement as he

THE CAMP AT HAWARA

broke from some covert at one end of the marsh and tried to slip away through some deep water. Three spears, of whom I was one, set off in pursuit. My pony, a young and untried animal, took me ahead of the others, and I was gaining on the boar when he turned, faced me, and decided to charge. It was too much for the nerves of my pony, who promptly swung round and fled. Mr. Goschen, of the British Legation, got a first spear as the pig went by him, which made the beast charge fiercely the remaining horseman, our Minister's (Sir Arthur Nicolson's) eldest son, who cleverly met the boar with his spear and killed him with one thrust; which was lucky, as we were all up to our horses' girths in mud and water. On Tuesday one boar was killed after we had run him for some

3

time in thick covert. Again Mr. Goschen got first spear. Another was wounded and escaped, and we had two or three good runs over the same thick country. On Wednesday, the mud-lake beat, we only killed one pig, but I enjoyed a very good run of over a mile at racing pace through the scrub after a small boar which we finally lost in a thick and high covert down by the sea. We ran two or three other pigs later, but lost them owing to the unrideable nature of the ground.

On Thursday the beat was the cork woods, and I was posted with some of the other spears along the edge of the marsh bordering the covert from which two half-grown boars broke. We followed,

THE SPEARS' MORNING START

but at that end the mud was so sticky that we could not catch either, and both escaped. It was then decided that a young Spaniard and myself should remain on the opposite side while the other spears went back to their places; so that should another pig break in the same place, we could go out to meet him and so lessen his chance of escape. After some time we saw three of our companions opposite riding a pig which had evidently broken in the middle of the beat. I soon perceived him, and noted his exceptional size—he looked enormous even from a long distance off; and I saw he was well ahead of his pursuers and was crossing some deep water in the middle of the marsh. Calling to my companion to follow, I galloped down along the edge of the marsh. When opposite the boar I halted. He did so too, at about seventy yards' distance, and we faced each other for a moment. I knew he was thinking whether to charge or not, and being an animal of prompt decision it did not take him long to make up his mind to do so. I turned my horse, who stood like a rock, broadside on, and, lowering my spear, waited for him. He came at a terrific pace with his head up and the water flying on either side of him as he dashed through it. My spear, however, met him full in the chest, and entered quite two feet, owing to the pace at which he was charging. It must have gone

NIZAM

right through his heart, for he staggered with the shock, fell over, and after one ineffectual attempt to rise, sank back in the rushes, and died.

It was all over before my companion and the other spears who had followed him across the marsh could get to me. The Spanish muleteer (whose duty it is to bring in the boars to camp, as the Mahomedan beaters will not touch the unclean beast) was ordered to fetch this veteran, and it was with the greatest excitement that I awaited his arrival, for, owing to the mud, water, and rushes where he fell, it was difficult to estimate his size. We were amazed, when he finally arrived, to see what a huge brute he was. We measured

X X 2

him carefully. He was 39 inches from wither to point of toe, and 6 feet 3 inches from snout to tip of tail. His tusks also were very fine, but I sent the whole head to England to be mounted without taking their measurements. A friend in camp took a photograph of myself with the boar, but owing to the failing light it did not turn out well; but I send one of my good horse Nizam, whose steadiness enabled me to make the lucky thrust which no doubt saved

COLONEL MANSEL-PLEYDELL, "MANAGER" OF THE TENT CLUB

me a very nasty conflict, and himself, most probably, from being badly wounded.

On the Saturday following (our last day's sport) this conviction was brought more forcibly home to me as a very big boar went away on the same marsh, and broke in much the same place, closely followed by two spears, Lord Cranley, of our legation, and Mr. Philip, the American Vice-Consul. As soon as the boar realized that he was being followed he stopped suddenly and charged Mr. Philip, who scored first spear, but at the sacrifice of a broken lance. The pig then again charged his horse, and inflicted a most severe wound on its hind-quarters, which laid the poor animal up for some months. Lord Cranley then rode at the pig, had his spear broken and his horse wounded, but luckily not so badly as the other had been. At this juncture I and others came up, and we all speared the pig, my spear being also splintered. Finally he retired into the marsh, evidently very sick indeed, and lay down, but got up when Captain Wilson of the 74th Highlanders gave the final thrust which ended the existence of this plucky but dangerous beast. He measured 34 inches, and was in full vigour of life and strength. The Belgian Minister and I enjoyed a pretty gallop across the drier end of the marsh a little later, and I succeeded in twice spearing a small boar which escaped wounded into the covert beyond, and we lost him.

This ended our week's sport. It was a more successful one than we had enjoyed for many seasons, a fact which gave much satisfaction to " the Manager," who takes a good deal of trouble, has all the worry, and, being most of the time with the beaters, has less chance of sport than any of the other spears ; so his gratification did not fail to increase that of the rest of the party.

INDIA N PIG-STICKING

BY R. D. RUDOLF

IT is early dawn in a Bengal April morning—so early, that the first glow of the rising sun has not yet touched the distant snowy peaks of the Himalayas. The moon is still high in the sky. But we have a long day's work ahead of us, so our servants had instructions last night to wake us at this very early hour, and, sleepy as they may be, they know well that if the sun should rise and we still in bed, he would shine a sad day for them. How easy it is the night before to gaily give the order that one is to be called at 4 A.M., and what a cruel thing that same order appears in its fulfilment! However, we have much to tempt us from our beds to-day, for are we not going to beat the famous Namiarali jungle for pig? and all who know anything of the district are aware that we have here a sure find, and that, at the present moment, many great boar are probably just getting to sleep in their lairs amongst the long grass, after a night of feasting in the villagers' fields of sweet potatoes near by, and of drinking from the river that flows past their retreat.

Presently all of us, twelve men in number, are dressed, booted and spurred, and sleepily sitting round a long table in the great verandah, doing our best to get outside of bacon and eggs and other eatables before starting. The jungle is some three miles distant and all our riding horses have already started, each led by his native groom, who carries over his shoulder a bundle of bag-spears. These are made with steel heads, of myrtle-leaf or diamond shape, mounted on male or solid bamboo shafts about seven feet long, and having, at the butt end, a leaden weight to balance the weapon.

Early breakfast does not take long, and soon we are driving in dog-carts to the scene of the sport. The sun is now just appearing and has no power as yet, and we drive bare-headed, in order thoroughly to enjoy the cool fresh air, scented with the odour of many tropical flowers. The dusty road winds through silent grooves of mango trees and past native villages, where the inhabitants are beginning to turn out sleepily to attend the drive, or to work in the fields, and groups of dusty little urchins stare at us with wide-open eyes, shrieking with delight at the unusual spectacle of so many sahibs.

Arriving at our destination we see, in a large mango wood, all the horses and ponies, and at a little distance a group of twenty-five or thirty elephants, surrounded by hundreds of natives. These men are the beaters, and for the sum of two annas, or less, a day they will tramp through the long grass and brushwood in rows, threshing about with their long bamboo sticks (called ' lahties '), shouting loudly and beating tom-toms to frighten the pig from their lairs. Few Europeans, bare-footed and almost unarmed, would care to join in this work at any salary, and yet these fellows seem to enjoy it and, as a rule, are not afraid. In nearly every drive some of them are hurt by pig, and, as a rule, it is the timid ones, who keep outside the jungle, that suffer. For a little while we walk about among the trees inspecting the horses and hearing stories of former struggles with pig, in which this or that animal saved his rider from danger, or, by his speed and handiness, gained for him the coveted 'first spear.' Presently our host drives up, genial and smiling, and all mount their horses, except a few who get on the elephants in order to direct the drive. The whole army of elephants, beaters, and horsemen make for the jungle near by. This jungle consists of a large patch of long grass, measuring perhaps a mile in length and half a mile in width and tapering at both ends. The grass is high enough completely to conceal the beaters and, in many places, the elephants, and is thickly studded with thorn-bushes, which, later in the day, play havoc with our clothing and skins. Our host, who is an old hand at the business, mounts a great elephant and takes his position in the centre of the line, into which all the elephants have been formed at one end of the jungle, while the beaters—who number upwards of a thousand men

—are with much noise at last got into line behind the elephants. The riders, each armed with a spear, are formed into two parties, which take up their positions at the sides of the jungle, being careful not to get ahead of the line. The drive then begins. What a noise the beaters make, as they shout, beat tom-toms, and whack the grass with their sticks! Flocks of little birds dart up here and there, startled by the strange sounds, and occasionally a partridge or quail goes whirring away to settle further on. Crash! goes some heavy animal there through the jungle ahead of the line, and the nervous novice, trembling with excitement,

OUT RUSHES A GREY BRISTLY OLD PIG

and eager to distinguish himself, grasps his spear and with bated breath watches the long grass wave, as the hidden fugitive makes his way to the edge of the jungle. He bounds into the open, to show himself, not a pig, but a neilghi, one of the largest known species of dear. He pauses for a moment, terrified by the sight of the horsemen, and then starts away at a long canter for a distant thicket. Ping, goes a rifle from a howdah on one of the elephants, and a spurt of dust just beyond him shows how narrow an escape he had. The horsemen, who are after pig alone, feel glad to see the beautiful creature reach a safe distance; but the natives think differently, as he and his numerous wives and family play havoc amongst their órops. Presently out sneaks a jackal, but neither spears nor bullets stoop to his level, and a few village dogs speed him to a safe retreat.

More neilghi break away, and our novice is almost despairing of seeing the great boar, of which he perchance dreamt last night, when suddenly, close to him, with a hoarse grunt, out rushes a grey bristly old pig, angry at being disturbed in his morning nap, and grunting vengeance on all who come near him. At a signal the line stops, and Mr. Pig, after having been given a start to get him free of the jungle, is hotly pursed by the party of that side. Away he goes at a lumbering canter, which looks slow until you try to catch him, and then you realise how fast he is really travelling. Straight for a distant jungle he heads, for, brave as he is, a boar generally prefers flight to battle, although he seldom fears the latter when he cannot comfortably escape. He has a clear mile to cover before he reaches the haven, and, ere he has accomplished half the distance, M——, our host's head assistant and right-hand man, is up to him on 'Robin,' his well-tried Arab. Just as he is about to plunge his spear well home, the pig suddenly alters his course, and goes off to the left across the horse as fast as ever. M—— is thus temporarily thrown out, and the second man, turning his horse in time, is on the pig's track. He is more fortunate, for the fugitive growing angrier as his wind gets shorter, suddenly turns in under the horse, meaning to rip him with his powerful tushes as he passes under him; but N——'s spear meets him in the shoulder, and over he rolls with the force of the shock. N—— has thus gained the first spear, and the head of the pig will belong to him when it is captured. But Mr. Pig is very far from consenting to this easy disposal of his cranium, the prod which he has received has only thoroughly aroused him, and, abandoning all thought of flight, he rushes wildly at the nearest horseman, again only to meet with the sharp point of a spear. After two or three such unsuccessful attacks, he stands still and glares at his enemies, the very picture of impotent fury and indomitable courage. Ride near him who may, he is sure to charge, but each time he gets the worst of it; then, with one grunt of anger he rolls over on his side, and is soon despatched. Dismounting, we gather round him and admire his fine tushes, which curl up on each side of his mouth, and with which he would so easily and so willingly have ripped up our horses or ourselves. Some natives come up and pluck the coarse bristles from his back, to be used by them as a medicine. He measures thirty-two inches

OVER HE ROLLS WITH THE FORCE OF THE SHOCK

at the shoulder, and is a fine beginning to the day's bag. N——makes a notch, or some such mark, in one of the ears, in order that he may recognise his property later on ; and, as we ride slowly back to the line of beaters, we can see a cart, drawn by bullocks, going leisurely out to bring in the carcass.

The beat continues, and, as the line nears the far end of the jungle, more and more of the inmates are driven out. There goes a whole herd of neilghi, making for the thin strip of wood along the river bank, and jackals by the dozen, snarling, find a safer refuge. The cry of 'Pig!' again goes up, and the line stops ; but he breaks back through them all, and gets a temporary retreat behind the line. The line has now reached the narrow end of the jungle, and the elephants are neared to each other until their sides touch and the line of beaters is several men deep. The grass in front of them is alive with all kinds of birds and beasts, and these fly and break away across the open. There is a crowd of yelping pariahs surrounding a surly bristling old porcupine, and here, just in front of us, a sow, followed by a row of little pigs, makes for the river bank unpursued ; for we do not wage war on women and children. The village dogs, however, have no such code of honour, and closely follow the family, until the indignant mother turns on them savagely, and thus covers the retreat of her funny looking offspring. The natives are unwilling to spare these young ones, and fling their sticks at them, until sternly reprimanded by the jemidars (native overseers). One or two more pig break out at the very end of the grass, but, on seeing what is awaiting them, turn and get back round the ends of the line of beaters. And woe betide any beater who, from faint-heartedness or sore feet, remains outside the edge of the grass, and thus gets in the way ; for the pig will make straight for him, and, as he passes under, will cut him with his sharp tushes. There is one man down already, and bleeding too. He is carried into the open, and, sure enough, has a nasty gash behind the knee. Luckily we have a doctor in the party, who, knowing what to expect, has not come unprepared. The line of elephants and beaters is now turned round, and slowly works over the same ground again ; but the three or four good boar which we know have broken back will not come out, but lie close in their lairs. The elephants warm to their work, beat the grass with their trunks, and push themselves through the thorn-bushes in a determined manner. Then one smells or sees a pig, and, with his trunk thrown up and trumpeting shrilly, rushes forward and tries to trample or kneel on the foe. But it is all no use, the pig will not break. The beaters and elephants are withdrawn.

Whew! how hot it is! and can it be possible that the sun, which earlier in the day looked so cool and harmless, is now

CUT HIM WITH HIS SHARP TUSHES

burning up everything? But it will very soon be hotter still, for dozens of natives are busy firing the grass; and soon a red line of flame can be seen advancing steadily through the jungle, fanned by the hot west wind, which daily rises about 9 A.M., and filling the air with black specks and smoke, leaving behind it a blackened and but partially burnt grass; for some of this is too green to burn well.

The pig do not like this form of attack, and presently one breaks away on the far side, and we watch the party yonder pursue him. He takes them over a nasty country, and when the foremost rider is up to him there are only three of the five who started in their saddles, two riderless horses being seen galloping away towards the distant factory, followed by their fleet-footed grooms. Neither rider seems to be hurt, and they are soon mounted again, probably vowing vengeance on the cause of their disaster. The pig is now at bay some two miles away, and soon the party riding back show that he has been added to our bag.

It is now nearly 10 A.M., and the heat is terrible, both from the sun and from the jungle fires. The air is full of smoke, and a thirst such as can only be equalled under similar circumstances takes hold upon us, so that we are right glad when our host dismisses the beaters and elephants, who make helter-skelter for the river, while we head for the comforting shade of a great pepul tree. Under it our clean white-clothed servants have spread a repast of cold viands. They help us to long iced drinks of various kinds ('shandy gaff' being the favourite).

What luxury it is, after the heat and hard work of the morning, to lie in the grateful shade of the grand old tree, and as the clouds of tobacco smoke curl upwards to gaze through them into the leafy depths, and dreamily picture the scene of perhaps a hundred years ago, when a widow was voluntarily burnt upon the funeral pyre of her husband, in the ashes of which this tree was planted to mark the spot! Here have come, year after year, the simple villagers, with their poor offerings of flowers and coins; for the tree is sacred to the memory of those whose tomb it marks, and weird spirits dwell amidst its branches who must be propitiated.

The scene around us is a truly Eastern one. In the centre are the white men in various postures, all trying to get the greatest amount of rest in the shortest time, and looking thoroughly disreputable in large solar topees, flannel shirts, and riding breeches, well begrimed with the smoke of the burning jungle. Can these be our spotlessly clean comrades of last night's dinner? Behind them are the table servants, busy with knives and forks, tumblers, &c. There are our horses, enjoying the shade and rest, and drinking greedily their *suttoo* (Indian meal and water), or sucking at pieces of ice, which they love on

T 2

15

such occasions. Yonder, beside the bullock carts, the drawers of which are lying down peacefully chewing the cud, lie the results of the morning's sport—two boar, looking grim and terrifying even in death. Around all is a brown circle of villagers, staring hard at the sahibs and their strange ways.

To horse once more, and now the old hands mount their best nags, for they know that the pig always break most freely in the heat of the day, and that probably the best sport is yet to come. The fierceness of the jungle fire is dying down, but several distant patches of grass are still in full blaze, and there, sure enough, are two pig, trotting back to the charred remains of their lairs, whilst a third is seen beyond the river, leisurely making across an open country. Our host points him out, with the remark that he is heading for a jungle full five miles away. Three of us ford the river and get on to his track, while the rest of the party follow the two first seen. The river runs shallow over the ford and the water is hot from the sun. The pig soon sees that he is pursued, and quickens his pace, but, after a fine gallop, we are up to him, and in the middle of an indigo field, the green crop in which is only a few inches high, he turns to bay at two or three village dogs who are snapping at his heels. H—— is up to him at once and sticks him in good style, but the pig charging at the same moment the spear comes out, and turning round, goes clean through the shoulder of H——'s horse. This is an accident which not infrequently happens, and may be a source of danger to a rider, and the tiro is always warned against lightly losing hold of his spear, although sometimes this cannot be helped. In this case, fortunately, the wound is only skin deep. The pig is now in grand fighting form, and charges anyone who approaches him. My pony is new at the work, and suddenly shying round at the most critical moment, nearly deposits her rider on the ground in front of the enemy ; but sheer urgency makes me strong, and I somehow or other clamber back from the region of her tail into the saddle, longing for my tried horse of the morning, who would not thus have played me false. Two spears have been left sticking in the pig, and as he moves these wag about and make it a very difficult matter to approach him. A lucky spear, delivered in the neck, however, suddenly drops him in his tracks, and he ceases to be a terror to the country around.

Back we hurry across the river, for we see our companions there galloping in parties of two and three in several directions, and know that there is grand fun going on. The river bank is lined with hundreds of natives, most of whom have come simply

NEARLY DEPOSITS HER RIDER ON THE GROUND IN FRONT OF THE ENEMY

to watch the sport, but a few are armed with rough home-made
spears, and are waiting there for any pig who may swim the
stream. Surely enough there is one doing so now, and as he
approaches the opposite bank these rascals wade out to meet
him and kill him in the water with no risk to themselves, for he
is powerless to hurt them when thus situated. This is poaching
of the worst kind, and the pig is taken from them, as are also the
spears, by the jemidars. Stern justice will be meted out to the
evil-doers at the factory to-morrow, for be it known that an
indigo-planter is a very autocrat in the country round his
factory, and of all sins perhaps our host will look with least
leniency on illicit pig murder.

The sport wears on, some other pieces of jungle are beaten and
burnt, and by 5 o'clock our bag has risen to nine boar and one
sow. The latter would not have been numbered among the slain
but that she wilfully charged our host, and he in self-defence laid
her low. Several natives have been cut, but none seriously, and
the carefully tended wounded are looked upon with almost
envious eyes by their brother beaters, for they will obtain good
food and lodging at our host's expense until their wounds are
healed, and may also receive backsheesh. This reminds one of the
native prisoner who, on the night of his dismissal from gaol,
committed a burglary upon that hospitable retreat in order that
he might again be shut up where good food and lodging were
so easily obtained from a munificent government.

The sun is nearing the western horizon and looks big and
red through the hot dusty air. The high hot west wind, which
has been blowing since late morning, is now falling, but everything is so heated up that it will be hours before the air will
begin to regain that refreshing coolness which is so characteristic
of an Indian early morning. The whole country looks desolate,
the burnt jungle and scorched trees lending a very bleak appearance to the scene. The elephants and beaters are tired out,
and so are we, when our host gives the now welcome signal to
stop the sport. The beaters and elephants at once make for the
river and drink of the hot polluted stream as if it were a cool
fountain. We are treated to something more to our taste, and
then, as the sun drops suddenly behind yonder sandhill, we get
into our dog-carts, and, tired, dirty, scratched, but happy, make
for the bungalow. Before we have driven the three miles, night
has fallen, pariah dogs bark at us from neighbouring villages,
and the jackals, grown bold in the darkness, howl in a dismal
manner. Through the now still air we hear the murmur of

many voices in the country round, for the population of this district is nearly 1,000 to the square mile, and the distant beating of a tom-tom away to the north declares a wedding ceremony in progress.

A bath, clean clothes, and cigarettes make new beings of us, and we stroll down to the stables and see the pig laid out, where they look big and terrible in the pale moonlight. After this, the head groom cuts off their heads, and the flesh is divided amongst the syces and grass-cuts, who will spend most of the night round great bonfires roasting, and feasting, and talking.

A good dinner under the cooling influence of great punkahs is a very grateful ceremony. It is very late before our host has replied to the drinking of his health—' with Highland honours '— and we sing ' Auld Lang Syne ' and retire to our beds to fall into dreamless sleep.

www.ingramcontent.com/pod-product-compliance
Lightning Source LLC
Chambersburg PA
CBHW031009090426
42737CB00008B/751